For Alexander

Island of the Mighty

by Haydn Middleton

illustrated by Anthea Toorchen

Oxford

DRAW closer to the fire now. The players are setting aside their harps, their pipes and fiddles. Outside in the stormy cold the sentries keep a constant watch. Come, warm your hands, and hear of how this Island used to be. Back before the darkness fell. Back through the forests of time. Back to the coming of the giants . . .

In the very beginning, the giants came to Wales. They had sailed from the Summer Country, far across the Hazy Sea. Hu Gadarn was their leader, Hu the Mighty. They came with the gifts of music and song, but without the arts of war. A noble race they were, tall as the trees and twice as broad. Their faces shone with wisdom we have long since lost. These giants had knowledge of the land, sea and stars. They knew where to build their circles of stone, knew where to carve their dragons in the hillsides. When they sat singing their beautiful songs, the waters and the grasses seemed to sing with them.

Island of the Mighty, that was how this land was known. And the giants made it bloom. Its trees bent low with their burdens of fruit, its waters teemed with rich harvests of fish. Island of the Mighty – a jewel of a land, girt on every side by heavy blue chains of sea. To the east, it was said, lay the lands of the Purple City. To the west lay something stranger – another island, sometimes there, sometimes not. Ireland, you might have called it. But others knew it as Tara, one of the realms of Anuvin. And no one had ever been there . . .

The time of the giants was a golden age indeed. Each generation melted away before the next, and the line of Hu the Mighty kept the Island's crown. The land and its lords moved in rare harmony. There seemed no reason why this harmony should ever cease – especially when the noblest giant of all became king.

His name was Bran. Bran the Blessed.

The Last of the Giant Kings

THE Island of the Mighty was in full flower. Its heartbeat was strong, its people knew nothing but peace and plenty. Year by year the grass grew greener, the rivers ran faster. Bran the Blessed had been king for longer than anyone could remember. He spent his days in tents or under the sheltering sky, for no house had ever been built that could contain him. Wherever he went, ravens came to him and perched on his awesome shoulders. He led a simple life, like all the giant kings before him, drawing his pleasure from the joy of his people.

One bright afternoon, Bran's court was at Harlech. While his people feasted and danced, Bran sat on a great rock by the coast, gazing out to

sea. Beside him was his sister, the lovely Branwen. Unlike Bran, she was not a giant. Throughout the Island now there were fewer giants than there had been before. But Bran valued his sister's advice, and she was always with him.

As the two of them sat in silence, the air grew thin and sharp. Suddenly a yellow ox-hide appeared on the rock beside them. Then a grim-faced man and woman were standing on the hide. The man was squat but looked very strong. His yellow-red hair was matted with blood, and he stank of hot iron. The woman was larger than he, and pregnant too. Both were dressed in filthy skins, and they dripped with weapons – axes and swords and the cruellest knives imaginable.

'Greetings,' said Bran, although he did not know them, 'You are welcome at my court.'

'Greetings, king,' the man replied. His voice was hoarse and his teeth were rotten. 'I am Lassar. This woman is my wife, Kymidei. We have come from the Lake of the Cauldron, in Anuvin, and we seek refuge in your Island. In return, we've brought you something that you need.' With that, he slung a great pearl-edged cauldron from his back and set it down before Bran the Blessed.

'I thank you,' said Bran, 'But I have cauldrons enough in my kitchens. Why should I need another?'

The man called Lassar smiled, and his wife cackled noisily. 'You have no cauldron like this one, king,' he said. 'This is the Cauldron of Rebirth. If you take a warrior who has been killed today and throw him inside, tomorrow he'll live again, and fight as well as ever, although he won't be able to speak.'

'But still,' said Bran, 'I have no need of such a marvellous thing. This Island is a place of peace. We have no wars here. We have no warriors.'

Lassar smiled a crooked smile at Kymidei, who began to cackle again. 'King,' he said slowly, 'times are changing. At this moment, your land is safe. But tell me this – what would happen if a force came here from one of the realms of Anuvin?'

At that point, a black-eyed man clambered up onto the rock. It was Efnisien, Bran the Blessed's brother. Efnisien was the one stain on the Island's honour. The sight of peace and friendship between men drove him to the verge of madness. Then he would unleash chaos like a wicked child untying a sackful of rats.

'Anuvin?' he barked at Lassar, 'What is this talk of Anuvin? Surely no such place exists?'

Lassar looked hard into Efnisien's eyes. 'The realms of Anuvin are as real as you or I,' he said.

'Then how can a man get there?' sneered Efnisien.

'There are many routes,' Lassar replied. 'Through a lake or a cave or a magical mist. Or sometimes by opening the third door.'

'But Anuvin is always changing,' croaked Kymidei, scratching at herself like a dog. 'It may take the shape of an island, a palace, a revolving tower of glass. One moment it is there before your eyes, the next it is nowhere at all!'

Lassar nodded his head, then turned back to Bran. 'To know of Anuvin is to beware of it,' he said. 'Once you know of such a place, it is unwise to be without warriors.' He rapped the Cauldron of Rebirth with his knuckles.

'And that is why this cauldron is well worth having.'

Bran stared at the two strangers. Branwen looked into the face of her giant brother, and she sensed that he had known all this before. Meanwhile Efnisien stood with his eyes on the sea, his mind swarming with new thoughts.

'I accept your gift,' said Bran in the end, 'And lodgings will be found for both of you in this Island. I bid you welcome.'

In the years that followed, Kymidei gave birth not once, not twice, but over a hundred times. And each time she took to her childbed, she produced a fully-armed warrior. Bran the Blessed quartered these warriors in every part of the Island, where they built great turreted forts, and instructed Bran's people in warfare.

It was a restless time. One by one the remaining giants wandered away from the court, and into the Island's remotest places. It was as if they knew that the golden age was drawing to its close. Bran watched them go, and vainly he tried to control his brother Efnisien. For all the talk of war had further warped Efnisien's mind. He rode about the Island, dressed in the fashions of battle, sowing havoc wherever he went. Until at last one day, to everyone's relief, Efnisien simply disappeared. He left word that he had gone in search of Tara, the island realm of Anuvin. But he made a mistake in going. For while he was away, it was Tara that came to the Island of the Mighty – in the form of a splendid battlefleet . . .

There were thirteen ships, and they appeared on the horizon as suddenly as a crack in a mirror. Bran saw them first. He was sitting on his rock at Harlech, with Branwen and Lassar at his side.

'These ships mean no harm,' said Lassar. 'See, a shield has been raised above each deck, with its tip pointing upwards. That is the sign of peace.'

The ships drew closer; on board the first one were the most dazzlingly white horses that anyone had ever seen.

The men on this ship were putting out boats, and rowing swiftly towards the shore. Bran rose to his full height, and called out, 'You are welcome in this Island. But where do you come from, and what do you want?'

'Greetings, king!' the men cried back. 'We come from the island of Tara, where the great Matholuch is king. He is with us, in the first ship. He has heard that your sister Branwen is the loveliest girl in the world, and he wishes to ask you for her hand in marriage.'

'In that case,' said Bran thoughtfully, 'He must feast with me tonight, when we will discuss the matter at length.'

The feast was a huge success. Bran's people mixed like long-lost brothers with the men of Tara. And Bran found King Matholuch to be a fine suitor for his sister's hand. When the feasting was over, the two kings went to inspect the white horses, the pride of the court of Tara.

'When one of these creatures is sick,' said Matholuch, 'I myself pass the night in the stables tending it. If anyone so much as raises his voice to one of them, I regard it as a personal insult. I hold these horses as dear as life itself.'

On the following morning, Bran held a council of his noblemen. At once they advised him that Branwen should marry King Matholuch, to set the seal on a glorious union between the two islands. Within days, a vast array of tents and pavilions had been set up at Aberffraw, on the Welsh isle of Anglesey. And there the royal wedding took place, to the delight of every single guest. Only one man could have spoiled such a celebration – Efnisien. But he was still away searching for Anuvin, so his seat at the feasting table remained empty.

Early the next day, while the wedding guests still slept, King Matholuch's grooms went to the stables. There they attended to the white horses, making them ready for Matholuch's morning inspection. As they worked, a harsh voice barked up at them from the seashore below.

'Whose horses are those?'

The voice belonged to evil Efnisien. He had returned, having failed to find Anuvin!

'They belong to Matholuch, King of Tara,' the grooms replied. 'He loves them like his own children.'

Efnisien screwed up his face, 'And what in the world are they doing here?'

'The king himself is here,' they replied, more timidly now. 'Last night he married the noble lady Branwen, and these are his horses. . .' And with that, they hurried away. But Efnisien clenched his fists, till the knuckles showed white. He moved closer to the magnificent horses, who tossed their heads nervously.

'So,' he muttered, 'Branwen is given away! How dare they do this? How dare they hold a council and give her away in my absence? I'm her *brother*! I'll show them, I'll teach them.'

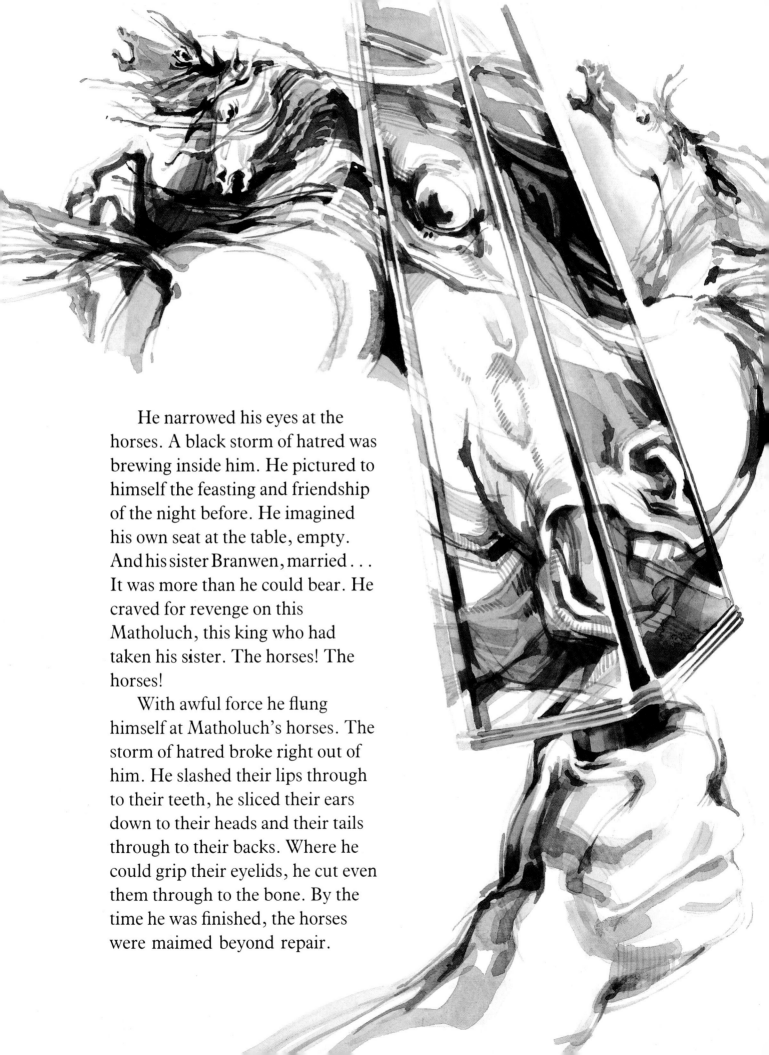

He narrowed his eyes at the horses. A black storm of hatred was brewing inside him. He pictured to himself the feasting and friendship of the night before. He imagined his own seat at the table, empty. And his sister Branwen, married . . . It was more than he could bear. He craved for revenge on this Matholuch, this king who had taken his sister. The horses! The horses!

With awful force he flung himself at Matholuch's horses. The storm of hatred broke right out of him. He slashed their lips through to their teeth, he sliced their ears down to their heads and their tails through to their backs. Where he could grip their eyelids, he cut even them through to the bone. By the time he was finished, the horses were maimed beyond repair.

Matholuch himself was the first to see the mutilations. He fell silent with grief, but his noblemen were fiercer. 'No king can put up with such an act of treachery!' they snarled. 'You must leave this land at once, this land which has treated you so cruelly!' Matholuch had to agree. Sadly he led his men back down to the ships.

But already Bran had learned of what had happened. 'This can only be Efnisien's work,' he said to his courtiers, and he sent three of them to Matholuch with this message:

'If you come back, I will give you one good horse for every one that was maimed. I will give you a silver staff as thick as your little finger and as tall as yourself. I will give you a plate of gold as broad as your face. If that is not enough, then come to me and name your own terms. But know this – Efnisien acted without my knowledge. He is a frantic man, stricken by some nameless evil. He is also my brother, so I cannot have him killed.'

Matholuch listened, paused for thought, then decided to accept Bran's generous offer. That night Bran laid on a special banquet. But it

was a listless and gloomy affair. Bran grew more determined than ever to make up for Efnisien's wickedness. 'Tomorrow,' he said to Matholuch at last, 'after the new horses and gifts have been delivered to you, I will give you one more present.'

'Yes?' said Matholuch. 'What will it be?'

'A wonderful cauldron. . .' And he went on to describe the magical powers of the Cauldron of Rebirth.

'I would be glad to accept such a cauldron,' Matholuch declared with a smile. 'Now let us lift this cloud of sorrow. After all, I have the greatest prize of all – the lovely Branwen.'

'Indeed you have,' said Bran. 'She will be missed in this Island. Look after her well. I could not bear to think of her unhappy.'

Two days later, the men of Tara left the Island with Branwen, the gifts and the Cauldron of Rebirth. Bran sat on his rock at Harlech with the ravens flapping on his shoulders, watching the thirteen handsome ships disappear into the distance. He was so silent that Lassar touched his arm. 'King,' he said, 'you seem so deep in thought. Are you not pleased by this union with the Island of Tara?'

'I am well pleased,' Bran replied. 'But I sense trouble ahead. Those maimed horses may still come between the two kingdoms.'

At that moment Efnisien climbed up onto the rock with them. 'King!' he cried, 'you look sad! Do you fear that our sister will be mistreated?'

Bran said nothing, but continued to look far out to sea.

Efnisien drew his sword and pointed it at the horizon. His eyes were sparkling with hate, his body was trembling with bloodlust. Bran felt the ravens leaving his shoulders, and he knew in his heart that disaster was going to follow.

The court at Tara was a sumptuous place. Branwen, the new queen, was greeted with great rejoicing. Noble visitors came from all over the Island bringing presents to show their loyalty. In return, Branwen gave each one a brooch or a ring or a treasured jewel. The people of Tara took her to their hearts, and Matholuch, like Bran before him, turned to her often for her wise advice. Then, before two years were up, Branwen gave birth to a baby boy, and they called him Gwern.

But that was the end of her happiness. For now Matholuch's nobles began to mutter again about the white horses – the horses that Branwen's brother had treated so savagely. The disgrace of it ate away at them like rust on an ancient shield. And soon they were baying for vengeance. One by one, they came to Matholuch and poisoned his mind against the Island of the Mighty. Finally Matholuch sent for Branwen. As soon as she saw the cold look in his eyes, she knew that something was wrong.

'Branwen,' he said, 'my noblemen cannot forget the affair of the white horses.'

Branwen's face went pale, but she replied with a steady voice, 'That may be so. But what can be done about it now?'

Matholuch looked away from her. 'They demand vengeance. It was your evil brother who inflicted this insult upon the court of Tara. Now it is you who must suffer. You are to lose your rights as queen.'

'I see,' said Branwen, trembling inside but showing no fear. 'And what will happen to me then, lord?'

'You will be stripped of your precious clothes and jewels. You will wear the rags of the lowest servant, and you will work in the royal kitchens. Now go.'

'I will go,' said Branwen, 'If that is your wish. But there is one thing you must never forget – I am the sister of Bran the Blessed. If he should hear of my suffering, he will shift heaven and earth to put an end to it.'

But for three heartbreaking years, Branwen knew nothing but misery. She had to share her food with the royal hounds, and sometimes she was forced to sleep beside the pigs. Her master was the court butcher, a vicious little man who beat her hard and often. He made her perform the most shameful tasks, day and night, night and day. And not once in those three years was she allowed to set eyes on her own child, Gwern.

But in spite of all her hardships, Branwen never wept, never complained, never lost patience. Even in her tattered clothes, even when her face was smeared with mud, underneath she was still a queen.

And still she had her old wisdom too. For even though her plight seemed hopeless, she worked out a way of contacting Bran the Blessed.

One day Branwen found a tiny starling, motherless and frightened, amid the refuse outside the kitchens. The butcher saw it too, and ordered Branwen to kill it. But instead she smuggled it into her own filthy quarters. There she reared the creature in secret, feeding it with titbits of food, loving it like a mother. Slowly then, with the greatest care, she taught the starling to understand words.

When the bird was ready, she began to tell it about the Island of the Mighty. Time and again she would whisper, 'The Island's king is my brother, Bran the Blessed. No house has ever been built that can contain him. He is the strongest and wisest and noblest king in the world ...' Then she would describe Bran's appearance in the tiniest detail so that, when the time was right, the bird would be able to recognize him.

Finally, one dark evening, Branwen took the starling from its hiding place under her stone workbench. 'Now it is time,' she said softly, and gently she pinned a little note to the underside of its wing. 'You are to be my messenger. This note I have written is for Bran's eyes only. Take it far, far away from this island. Take it safely across the Hazy Sea, and, if there is any such thing as justice, you will come to the Island of the Mighty. Then you must seek out Bran the Blessed. I am sure that Bran

15

will know how to find me. He knows so much. He has the knowledge of the giants.'

Branwen kissed the starling lightly. 'May the stars be your guides and the winds give you speed,' she murmured, before tossing it up through the kitchen window, and out into the dark night. And then she began to wait. But hardly any time at all passed before a swineherd came rushing into the court of Matholuch, his face wide open with terror. 'What is it?' asked Matholuch. 'What news do you have?'

'Wondrous news, king,' gasped the swineherd. 'I was down on the coast, gazing out to sea, when suddenly I saw a forest moving towards me on the water!'

'A forest?' cried Matholuch. 'Did you see anything else?'

'Lord, I did. Beside the forest was a mighty mountain! And on the mountaintop there was a lofty ridge! And on each side of the ridge was a lake! And all these things were moving steadily towards the Island of Tara!'

Matholuch had no idea what any of this could mean. His noblemen were equally bewildered. 'Only one person in this court can explain such a wondrous sight,' said Matholuch. 'Fetch Branwen from the kitchens!'

Branwen came, and she listened as the swineherd repeated his story. 'Lady,' said Matholuch then, 'Tell us the meaning of these strange things.'

'Lord,' Branwen replied with a calm smile, 'I am no lady now. But the explanation is simple: the men of the Island of the Mighty have heard of my treatment here, and they are coming. Can you not see? The forest on the water is the masts and yardarms of their ships. The moving mountain is my brother, Bran the Blessed, wading through the sea. The lofty ridge is my brother's nose.'

'And the lakes?' asked Matholuch, stiffening with fear.

'My brother's eyes,' cried Branwen, 'black with fury at you and your land!'

No longer could she hide her delight. The starling had succeeded in its mission. Bran was on his way. At last her sorrows were coming to an end!

But Matholuch immediately called his noblemen about him. 'This is your doing!' he shouted at them. 'Vengeance has bred vengeance. What can we possibly do now? No army of ours can defeat Bran's force. No walled city could ever keep them out. They'll destroy this whole island! You must come up with a plan – and quickly!'

'King,' they said in reply, 'There is no cause for alarm. What we must do is this – retreat across the great River Liffey, then hack down all the bridges after us. There on the other side we'll be safe from Bran and his force. No swimmer has ever been able to cross the river, even at its narrowest point. The current is too strong, the waters too thick and black.'

'Good,' said Matholuch, although he looked uncertain. 'We will cross the river at once. And you, lady,' he said to Branwen, 'will come with us.'

'As you wish,' Branwen replied with a smile. 'But you will not escape from Bran the Blessed. I can promise you that.'

Bran's fleet landed on the coast of Tara. At once the march inland began. Helmets, spears, shields and trumpets flashed in the hot sun. It was a gigantic force. Very few men had stayed behind in the Island of the Mighty. Bran had chosen Lassar and his own son Caradoc to rule while he was gone. But as Bran strode out ahead of his men, Efnisien trotted delightedly by his side. Efnisien could barely wait for the killing and wounding to begin. His eyes were gleaming as keenly as his weapons.

Then they came to the great River Liffey, and Efnisien's spirits sank. 'Brother, look!' he yelled, 'there are the men of Tara, beyond the far bank. But they have wrecked all the bridges, and no man could hope to swim across this torrent of a river!' He glowered at the rearing, crashing black waters, thick with slime, deafeningly thunderous.

But Bran looked across the river, and there beyond the far bank he saw Branwen, her arms held by Matholuch. He smiled grimly, and then he declared, 'He who would be leader, let him be a bridge for his people!' And with that, he laid himself down in the mighty waters. With his arms outstretched, he grabbed hold of the far bank, making himself into the sturdiest of bridges. The mouths of the men of Tara fell open in amazement. They could only watch in silence as, amid a fanfare of trumpets, Bran's force crossed the river towards them.

At last Bran lifted himself out of the water. Not a man or a horse or a wagon had been lost. But even before he could shake himself dry, a band of Matholuch's noblemen approached him.

'Greetings, King!' they cried, 'welcome to Tara!' Our king swears that he means you nothing but good.'

'Is that so?' said Bran. 'Then how does he intend to pay for the wrong he has done to my sister Branwen?'

'This is his offer,' they replied at once. 'He will stand down as king.

He will pass the crown gladly to Branwen's son. Your nephew Gwern will be crowned King of Tara!'

Bran looked at Branwen again. She was much closer now, and in front of her he could see the boy Gwern. 'It is not enough,' he said to the noblemen. 'Your king must make a better offer. And he must make it before this river water dries on my body!'

The noblemen scuttled back to where Matholuch stood. For almost a minute the two armies faced each other, expecting the order for battle at any moment. Evil Efnisien stood beside Bran, thirsty for blood, swallowing eagerly and often. Then Matholuch's noblemen came back to Bran.

'Do you bring another offer?' asked Bran.

'King, we do. It is said that no house has ever been built that can contain you. Is that so?'

'It is. What of it?'

'Our lord Matholuch promises to build a soaring fortress. Not only will *you* be able to fit inside, one half of it will contain the force from the Island of the Mighty. The other half will contain the force of Tara. Then, when all are inside the fortress, Matholuch will offer the crown of Tara to Gwern – and we will all do homage to you as overlord.'

Bran began to nod gently. But before he could accept the offer, Efnisien darted in front of him.

'King!' he almost screamed, 'this is nothing but a foul trick. These people are cowards and liars. Tell them that the time for offers has passed. And let the fighting begin!'

Efnisien's words carried as far a Branwen. She wrenched herself free from Matholuch's grip, rushed forward and threw herself at Bran's feet. 'Brother, brother,' she begged. 'Don't listen to him! There must be no war over me. If this land is laid to waste, it would be no good to anyone – winner or loser. Trust Matholuch now. Accept his offer and let that be an end to it!'

Bran bent down and lifted his sister to her feet. He had never been able to ignore her advice. Solemnly he gave his acceptance to the noblemen, and a truce was agreed immediately.

Efnisien stalked away along the riverbank. He sheathed his sword in disgust. So there was to be no fighting after all. At least, not that day.

But still he had the taste for blood in his mouth. And he was convinced that the men of Tara were laying a trap . . .

For six days, the force from the Island of the Mighty sat watching as the fortress took shape. By the seventh morning it was finished. The first part of the bargain had been completed. Bran the Blessed paced around the walls very slowly. Never before had he stood before a building that he could enter. It covered the same area as a whole town. Its turrets were so high that they seemed to pierce the clouds.

The time came for Bran to lead his force inside. But as he set foot on the threshold, he found Efnisien barring his way. 'King,' said the devilish man, 'Let me go first.'

Bran smiled at him. 'You have no trust, Efnisien,' he said. 'Do you still suspect trickery of some kind?'

'I do indeed!' shouted Efnisien, loud enough for the men of Tara to hear. 'Let me inspect this fortress before you step inside. What is there to lose?'

Bran nodded his head, and Efnisien made his way into the great hall. He had never seen such grandeur. The workmanship was breathtaking, the coloured woods and stones dazzled the eye. It was as quiet as a cathedral, empty as an eggshell. One of the men of Tara came to Efnisien's side while he scanned the entire hall.

Efnisien tilted back his head to look at the ceiling. It was decorated by stars, and held in place by two hundred pillars of granite. But he noticed that a leather bag had been pegged to one of the pillars. Glancing around, he saw that every single pillar had a similar bag pegged to it.

He became deeply suspicious – and rightly so. For inside each bag, the men of Tara had placed a fully-armed warrior! As soon as the force from the Island of the Mighty was inside the hall, these warriors had been instructed to leap down and cut the guests to shreds.

'What is in this bag?' asked Efnisien, pointing at the nearest one.

'Why,' replied the man, looking shocked, 'only flour, friend.'

'In that case,' grinned Efnisien, reaching up and fumbling inside the bag, 'I can run my fingers through it!' And when his fingers felt a head

inside the bag, he began to squeeze. His fingers were like rods of iron. He squeezed and squeezed until his fingertips crashed through the skull and into the softness of the brain. Smiling broadly, he moved on to the next pillar. 'And what is in here?' he asked, pointing at the bag pegged there.

'Again, nothing but flour.' The man dared not tell the truth about the trap.

So Efnisien reached up again, found the warrior's head and crushed it between his fingers. On and on he went until he had killed one hundred and ninety-nine warriors. When he came to the last bag of all, he killed the two-hundredth warrior just as he had killed all the others, even though this one had a helmet on his head. Then Efnisien turned to the man at his side, and roared with laughter.

But the two armies, waiting outside the fortress, had grown impatient. Now they began to stream inside, marvelling at the magnificent hall. Things were moving very fast. Efnisien could not get close enough to Bran to tell him about the trap. All the men took up their positions in the two halves of the hall. Then, after prayers and fanfares, Bran and Matholuch signed a peace treaty. They swore that never again would there be bad feeling between the islands. And finally Matholuch himself placed the crown of Tara on Gwern's head.

When the crowning was over, Gwern ordered the men of Tara to go down on one knee to Bran, their overlord. Now all the parts of Matholuch's promise had been fulfilled – so the feasting could begin.

Efnisien sat slightly apart from the rest. His mood was fierce. The sight of so many peaceful, friendly faces wounded him as badly as a thousand sword-thrusts. Through narrowed eyes he watched the boy Gwern playing happily with Bran and his other uncles from Tara. Each uncle in turn gave the boy his blessing. And deep inside Efnisien, the black storm of hatred swelled again.

'That child is the son of my sister,' he growled. 'Why does he not come to me?'

Bran turned to him. 'What is troubling you, brother?' he asked.

'I said why does Gwern not come to me? He goes to his other uncles. Does he think me a worse man than them?' Efnisien had clenched his fists, the knuckles were showing white and raw.

'Gwern,' said Bran, 'Go to your uncle Efnisien. Receive his blessing.' The boy ran smiling down the hall. Efnisien hardly saw him coming. All he could see or hear or smell was hatred. He had been sucked right inside his storm of hatred. Nothing but disaster could satisfy him now, nothing but death and blood and wailing. The boy! The boy!

Efnisien launched himself out of his seat, seized Gwern by the ankles, and before anyone under that starry roof could stop him, he hurled the boy headlong into the fire!

Branwen screamed as her son burst into flames. She tried to leap after him. But it was too late, and Bran reached out and grabbed her with one of his huge hands. All around him, the men of both armies were rising in an uproar. The men of Tara swore to avenge the death of their boy-king. Every man reached for his weapon, screaming, cursing – the dreadful din that comes before slaughter. But to the ears of Evil Efnisien, standing on the table with his sword unsheathed, it sounded like the sweetest music in the world. At last he was going to slake his thirst for blood!

The battle began there in the marvellous hall.

'Let the Banner of the Mighty be raised,' boomed Bran. 'Let the fighting take its course!' And the great Banner which showed a single raven was unfurled, then hoisted where Bran stood. The giant king's face was knotted with bitterness. This is how it is with wars – anger breeds anger just as evil breeds evil. Efnisien saw that look on Bran's face. It was a look he had never expected to see on a giant. The sight of it, for some reason, troubled him . . .

Then Bran, protecting Branwen beneath his shield, hurled himself into the fighting.

Soon the floor was knee-deep in dead warriors. The battle spread outside the fortress, across fields and mountains, through the towns and villages of Tara. But back inside the fortress, the tide was turning. 'The Cauldron of Rebirth!' Matholuch had cried, 'Bring the Cauldron of Rebirth! Let us give new life to the fallen!' And Bran's men watched, aghast, as Matholuch's noblemen dragged the cauldron into the hall, then placed it above the fire. If only Bran had not given it away to Matholuch!

As soon as a man of Tara was killed, he was thrown inside. In no time the Cauldron of Rebirth was full. On the next morning the dead men sprang out with fresh life, brandishing swords, eager to dive back into the hellish war.

So it went on, for days, perhaps weeks. No one could keep track of the time. The smoke of destruction formed heavy black clouds over the land. The sun's rays were blocked out. Perpetual night closed in. Bran's men fought as ferociously as any army had ever fought. But they had no answer to the Cauldron of Rebirth. Day by day, their own numbers grew smaller and wearier. And day by day they were faced by fresh hordes from Tara. Bran the Blessed looked about him in fury. He knew

that his warriors had no hope of victory. Then, with a tormented roar that shook the fortress walls, he erupted into a frenzy of butchery.

Efnisien, in combat nearby, watched his giant brother in horror. He flung down his own sword, and staggered back from the demonic Bran. 'What is this evil that I've done?' Efnisien whispered hoarsely to himself, 'What have I *done*!' He didn't dare to believe what he was seeing. But it was true, desperately true – the good and gentle Bran, noblest of all kings, had become no more than a murderous brute!

The sight shocked Efnisien to the core. Tears began to spill down his blood-smeared cheeks. He fell to his knees, crushed by the weight of his own guilt. At last, the black storm of hatred inside him had blown itself out.

But he knew that he had to act. He had to make some kind of amends. He owed it to his king, his sister, his countrymen. They had all suffered so much because of his vile ways. His eyes fell on the cauldron.

Quickly he slipped himself in among the piles of dead warriors from Tara. Soon two noblemen came, and, believing him to be a man of Tara, tossed him inside the cauldron. Efnisien knew what he had to do.

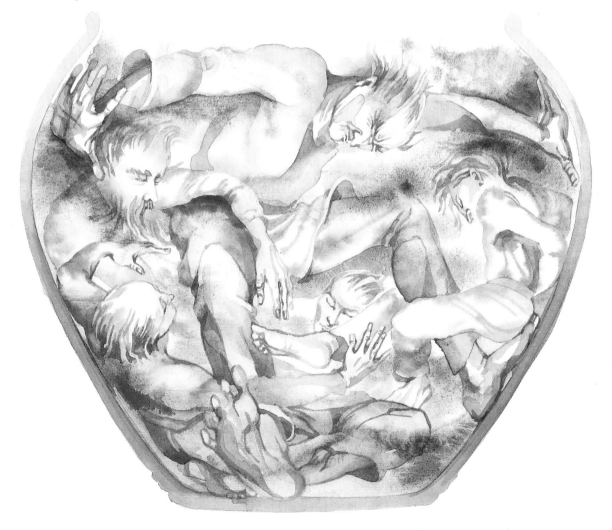

He stretched out his arms, then his legs, as far as they would go. He drew in two great lungfuls of air. Then with all the strength in his body, he pressed against the cauldron's sides. He pushed and strained, till every muscle quivered on the point of exhaustion. But just before he breathed out, the Cauldron of Rebirth cracked and burst apart into four great pieces.

And that was the end of Efnisien too. One good deed was not enough. The agony of his shame broke his heart, the agony of all his efforts broke his body. But he had done well to shatter the Cauldron of Rebirth. Had he not done so, every single member of Bran's force would have perished in the end. As it was, when the war finally ran its course, only fourteen souls remained alive in the whole of Tara.

Peace fell on the ruined land like mist over a graveyard. Not one man of Tara had survived; there were just five pregnant women, shivering together in a cave in the wilderness. On the side of the Island of the Mighty, death had spared only nine. One of the nine was Branwen, another was Bran himself. But the evil was not finished yet. For a poisoned spear had entered Bran's foot. And already the poison was swimming into the rest of his enormous body.

So Bran, Branwen and the seven others headed for the coast of Tara in silence. They left behind them the vast fortress, black against the broken sky, leaving behind the four great pieces of the Cauldron of Rebirth. They picked their way through the wasteland. And when they reached their ships at last, they made ready for their voyage back home.

At that point, Bran the Blessed spoke. 'Not one of these ships can carry me back,' he said, 'and the poison has made me too weak to wade. Yet I must go with you, to guide you back to the Island of the Mighty.'

Branwen looked up at him in despair. 'Then what are we to do, brother?'

'It is my wish,' said Bran steadily, 'that you cut off my head.'

'King!' cried the seven men in panic. 'What are you saying?'

'It is my wish,' Bran repeated. 'It is also my command. This is how it must be. I know what I know. A shadow is already falling across the Island of the Mighty. Listen now to what I have to say – and listen well.'

Branwen and the seven men looked helplessly at one another.

Silently they were shedding tears. But they knew that they had to obey their king – even if it meant cutting off his head.

So Bran continued, 'My head will guide you back safely to the Island of the Mighty. You will find the Island greatly changed. But you must take my head to the place called the White Mount, in the city of London. And there, along with the Banner of the Mighty, you must bury it, with my face turned to the south. Do you understand?'

The others nodded, numbed by all that had happened and all that was going to happen.

Then Bran continued to speak: 'Because of our war with Tara, the realms of Anuvin are now our enemies. In whatever way they can, they will seek to destroy the Island of the Mighty. But my head will combat the curse of Anuvin. Its powers will keep at bay all invaders from those realms. The Island will survive, and then will come the fourscore years of forgetfulness. But the people of the Island must always remember this – one day a palace will appear on the coast of Wales. Its hall will have three doors. Do not open the third door. Is that clear to you? *Do not open the third door.*'

The men were too bewildered to answer him. Bran smiled at them, then said:

'Throughout the voyage, and while you are on the road, my head will be as good a companion to you as I ever was. As for my body, you must bury it in this alder grove. Now the time has come. The deed must be done.'

Branwen turned away. But the seven men roused themselves, and severed Bran's head skilfully from his body. Just as Bran had said, the massive head was still able to speak and smile. Within days, it directed their ship within sight of Anglesey, off the coast of north Wales. When they reached the shore, they sat down to rest. And soon a group of ragged men and women passed by.

'Good people!' cried Branwen, 'Do you have news?'

'No,' they replied. 'There is no news any more.'

'But what has been happening here since Bran left for Tara?'

'A shadow has fallen across the Island – the shadow of war. Bran's son Caradoc has been overthrown. Lassar has disappeared. There is no peace any more. Only greed and noise and bloodshed.'

Branwen watched them move on and her face was like stone. Far across the waves lay the wasteland of Tara. And now she found the Island of the Mighty in turmoil too. 'Oh why was I ever born!' she moaned. 'Two good peoples have been ruined – and all because of me!' And the grief was too much for her to bear. There, at the point where the land met the sea, her heart simply broke inside her. The seven men carried her lifeless body to the bank of the River Alaw. They dug for her a four-sided grave, and then, unable to utter a word, they buried her.

At once they set out on their journey to London. When at last they reached the White Mount, they made a great hollow in its side. Tearfully they laid Bran the Blessed's noble head in the hollow. Then next to it they placed the Banner of the Mighty. As soon as they had covered both head and banner with earth, a great flock of silent ravens appeared in the sky overhead. And when the men had departed, the ravens alighted on the White Mount and looked about them.

Bran the Blessed was dead. The time of the giants had passed. But deep inside the White Mount, Bran's head continued to work its wonders.

How did it do so?

Bring more fuel for the fire. The night closes in and the cold wraps us all. Pass round the drinking horn, then listen. Listen to the next part of this tale. Years and years and years slipped by. Bran's head lay open-eyed in the White Mount, until men stopped knowing that it was there. And then came the three great plagues from Anuvin.

Young King Lud and the Three Great Plagues

THE people of the Island of the Mighty had a saying: 'Bad things happen in threes.' Young King Lud found out how true that was! Soon after he came to the throne, not one, not two, but three plagues struck the Island. And it all began when the Corannies arrived from across the Hazy Sea.

The Corannies were tiny folk, the size of a thumbnail and the colour of autumn leaves. They came at the dead of night, in eggshell boats lit by single candles. Only a handful of Island people were awake when the Corannies first appeared off the north coast. They saw the sea smothered in pinpricks of light, then watched as the little people jumped ashore.

'They look harmless enough,' said one watching northerner, 'And besides, they're so small that there will be plenty of room on this Island for both them and us.' But he was terribly wrong. Within several months – when it was already too late – the truth about the Corannies came out: they were water-tamers.

Water-tamers know how to make water work for them. Ordinary people can use wood and stone and the fertile soil, but they do not know the mighty secrets of water. The Corannies were different. They knew how to catch the rain before a drop touched the ground, then turn it into marvellous food and drink. They knew how to tie knots in the fastest-flowing streams, and weave the knotted water into gorgeous cloth. They could make bricks out of lakewater and build lovely shimmering cities.

They could even conjure messengers out of the moisture in the air, to bring them reports of conversations up to a thousand miles away!

But water-tamers use up huge amounts of water. Within three years of their arrival in the Island, the Corannies turned the whole northern region into a desert. The northerners had no idea what to do. Six times they gathered together and laid plans to stamp out the Corannies in surprise attacks. Six times the moisture-messengers warned the Corannies in advance – and six times their attacks ended in wretched failure.

Finally the northerners decided to travel south, to see King Lud at his court in the city of London. Perhaps *he* would know how to deal with the Corannies. But young Lud was scarcely more than a boy. He still needed time to learn the wisdom of kings.

As the northerners entered the middle region, a throng of crippled midlanders limped towards them. 'Go back, go back!' the midlanders screamed. 'This place is full of danger!'

'Can you tell us why?' said the northerners. And they listened in horror as the midlanders described the plague that had struck their region. It came as a scream in the night. Not the scream of any man or beast. More like an echo from the time of chaos, before the world was formed, when land and sea and sky were all one. Every May Eve it worked its way loose from the depths of the earth, then rose up and did its dreadful damage. With each May Eve it swept over a wider area, sucking the sense and strength from every living thing. Old men who heard it went mad. Women lost their beauty and their babies died inside them. Children were struck dumb, healthy men withered on the spot, fish, cows and flowers folded and died. There seemed to be no escape from the scream, and certainly no way of stopping it. The northerners looked at the midlanders with pity, but then marched on into the

southern region. They entered king Lud's court, bowed low, and told him about the Corannies and the scream.

Lud listened closely, and when they had finished he nodded. 'These things are said to happen in threes,' he said grimly.

'King, what do you mean?' asked the northerners, puzzled. 'What is the third plague?'

'An odd one indeed,' Lud replied. 'And it is laying waste the whole of the southern region. This is what happens: every night the strongest storehouses are raided, and emptied of their food supplies. Most of the people in the south are starving. Even here at my court in London we are growing famished.'

'Can you not post sentries outside the storehouses?' asked the northerners. 'Then you could catch the villainous thieves.'

Lud sighed. 'That is more difficult than it sounds. You see,

throughout the entire region, no one can keep awake during the night. But trust me; return to the north now, and I will do whatever I can for all my people. What more can I say to you?'

Lud racked his brains but he could think of no way of dealing with the plagues. Desperately he sought the help of the Island's cleverest wizards – the Druids. But even the Druids could suggest nothing. 'These are not ordinary plagues,' they told the king. 'In our opinion, they have been unleashed from the realms of Anuvin.'

'Anuvin?' said Lud quietly. 'How can I fight the forces of Anuvin, alone?'

That night, Lud wandered away from his palace in the lowest of spirits. He found that his footsteps were taking him in the direction of the White Mount. It was covered, as usual, by the flock of silent ravens. But there was neither man nor woman anywhere in sight. So Lud was shocked when a rich voice sounded clearly in his ear. 'Lud,' it said, 'Make at once for the northernmost point in Wales. From there take a boat to the isle of Anglesey. A wise man lives there in the woods. He has knowledge of the land and the sea and the stars. He will tell you how to drive out the three plagues.'

'Who are you?' cried Lud. 'Show yourself!' But he heard and saw nothing more. Lud did not know it, but that voice had belonged to Bran the Blessed. Deep inside the White Mount, the old king's head was still watching, listening, guarding.

Lud arrived at the northernmost point before dawn. A boat was waiting to take him across to Anglesey. Soon after they set off, thick mists came down over the water. The captain called out to Lud that his boat seemed to be guiding itself. Then the silhouette of a mysterious vessel appeared close by.

A crouched old man, swathed in rags and mistletoe, was standing on its deck. 'King!' he called out to Lud in a thin voice, 'I am the wise man of the woods that you seek. Come across to my ship, and I will talk to you of plagues.'

Lud waited until the ship was right alongside, then leapt across into it. To his amazement, he found that the entire ship was made of blue-green glass! The old man was standing before him, with a tray full of strange objects at his side.

'Old man,' said Lud, 'I thank you for meeting me. But who. . . who are you?'

The old man closed his eyes and laughed a creaky laugh. 'Who am I?' he said. 'Who am I? Let this be my answer:

Primary chief bard am I to Elphin,
And my original country is in the region of the summer stars;
Idno and Heinin called me Merlin,
One day every king will call me Taliesin.
I have been many shapes
Before I came to be like this.
I have been a sword in the hand.
I have been a shield in the fight.
I have been the string of a harp.
I have been called Lassar,
Father of the fully-armed men.
I shall be until the day of doom upon the face of the earth!'

Lud had no answer to any of this. All he could do was stare at the wise
man's rotten teeth, and smell his curious smell of hot iron. Then the
withered old creature took a long bronze horn from the tray. He placed
the wide end against Lud's ear. Then he raised the other to his own lips
and said, 'When we speak of the Corannies, we must use this horn. It

35

will stop their moisture-messengers from snatching our words. Are you ready for me to begin?'

Lud nodded.

'Then this is what you must do . . .'

Lud listened hard, but all he could hear was scratching and panting. He listened harder, and when at last he heard the wise man's words, he backed away, amazed and upset. For the wise man had said:

'Poor King Lud, poor King Lud!

Like a cow he chews the cud!'

'What troubles you?' asked the wise man, lowering the horn.

'What you said to me,' said Lud, badly offended. 'It was outrageous!'

The wise man frowned, but a smile was creeping down his face. 'Come back here,' he told Lud. 'Put the horn to your ear again and listen.'

Lud did as he was told, but not very willingly. The wise man was speaking again. Lud heard more scratching and panting, then these words:

'Lud is dull, Lud is dull!

Throw him over the side of the hull!'

Lud pushed the horn from his ear, too disgusted to speak.

'Did my words hurt you?' asked the wise man with a grin. But Lud simply scowled back, cursing himself for travelling all this way just to be insulted. 'Come now,' wheezed the wise man, 'stop looking so fierce. The truth of this is simple. The Corannies have heard that you are here. They have sent a devil to hide inside the horn. Whenever I start to tell you what you must do, the devil catches my words and twists them into insults. These Corannies are cunning creatures. After all, they come from Anuvin.'

'But what do we do now?' asked Lud, relieved.

'This,' said the wise man, lifting a flagon of wine from the tray, and handing the horn to Lud. The young king held it upright, with its mouthpiece on the shining deck. The wise man began to pour in the wine. For several minutes nothing happened. But just as the thick red liquid was reaching the brim, a high-pitched gurgling squeak came from inside the horn. Suddenly a spoon-sized devil with flap ears burst up through the surface. It was holding its nose in terror, and splashing for dear life.

The wise man flicked out his hand, plucked the devil by one ear and tossed the creature overboard. 'Now at last we can talk,' he said. And then he told Lud all that he needed to know.

Lud felt more anxious than ever. He now knew how to deal with the Corannies. But the wise man had also told him where the scream came from – and who was stealing the food. In both cases, this new knowledge terrified him.

Dawn was breaking, and the ship of glass took Lud right back to the mainland of Wales.

'Remember,' said the wise man, 'only a king can deal with plagues such as these. Follow my instructions, but tell no one what you are doing. Now take these . . .' He pressed a finger to his lips, then handed Lud from his tray a bagful of dead blue insects from Anuvin, a bronze cauldron large enough to contain a man, a silken sheet the colour of mud, and two stone jars. Lud understood how he was meant to use them all, and he nodded his thanks. But the last words he heard from the wise man were dark and

strange. 'If you see the third door,' he called out from his ship of glass, 'if ever you see the third door, do not open it . . .'

As soon as Lud was back in London he set to work. Standing alone outside his palace, he made this announcement: 'Corannies, although you are far away I know that you can hear me. Listen to my offer. I want my people to live in peace with you. We must put an end to the hatred between us. So come, as honoured guests, to my court here in London. Do not be afraid. I swear to do nothing to you that I would not do also to my own people . . .'

The Corannies arrived outside the palace gates early the next morning. To announce their arrival, they built a palace of their own out of the morning dew. Lud spoke to them from inside his own chamber. 'Welcome,' he said. 'Please proceed to the great hall. My own people have already assembled there. Soon I will be with you.' The moisture-messengers took his words to the Corannies. At once they began to stream into the palace. Lud knew that he had to act quickly.

He took the bronze cauldron that the old man had given him. He filled it with water, then tipped in the dead blue insects from Anuvin. After mashing them for a few moments, the water turned a deep ocean blue.

Taking in two great lungfuls of air, he heaved the cauldron off the ground. Without daring to breathe out, he hauled it up the stairway to the minstrels' gallery, high above the great hall. He looked down at the swarms of Corannies, mingling unharmed with his own bewildered people. And before a single pair of eyes could look up, he lifted the cauldron onto the balcony, then tilted it so that the blue mixture cascaded over them all!

A fearful cry went up from his people – and a steeper, strangled moan came from the Corannies. Everyone had turned a beautiful shade of blue. 'But look at the Corannies!' screamed one of the ladies of the court. They were indeed worth looking at. The blue-stained water-tamers had huddled into an enormous pile. It reached almost as high as Lud's balcony, a seething mass of blue, humming a single note of anguish. And it was transforming itself into – of all things – a pyramid!

There it was, finished now, a perfect silent pyramid, made of what looked like rippling water. It stood for less than a minute. A great shudder passed through it from base to tip. Then it reared up from the ground, and collapsed like a monstrous wave over all the awestruck courtiers.

That was the end of it. Every last drop of the pyramid disappeared, leaving Lud's people miraculously dry – and clean! Not one of the Corannies had survived. The people raised a cheer to their quick-witted king. But Lud was no longer in the hall to hear them.

Instead he was driving a cart towards the middle region. He had no time to lose, for this was May Eve, the day of the ghastly scream. Swiftly he headed for the mid-point of the Island. And in the cart with him were the cauldron, the silken cloth the colour of mud, the two stone jars, and gallons and gallons of the sweetest, sickliest wine in the Island.

Late in the afternoon he reached his destination. A flat and foggy place it was, far from any signs of life. There was a large murky lake, and, nearby, a small circle of standing stones. It was all exactly as the wise man had said.

Carefully Lud wedged the cauldron into the thick mud at the lake's edge. He filled it with the wine, then spread the silken cloth across it. In

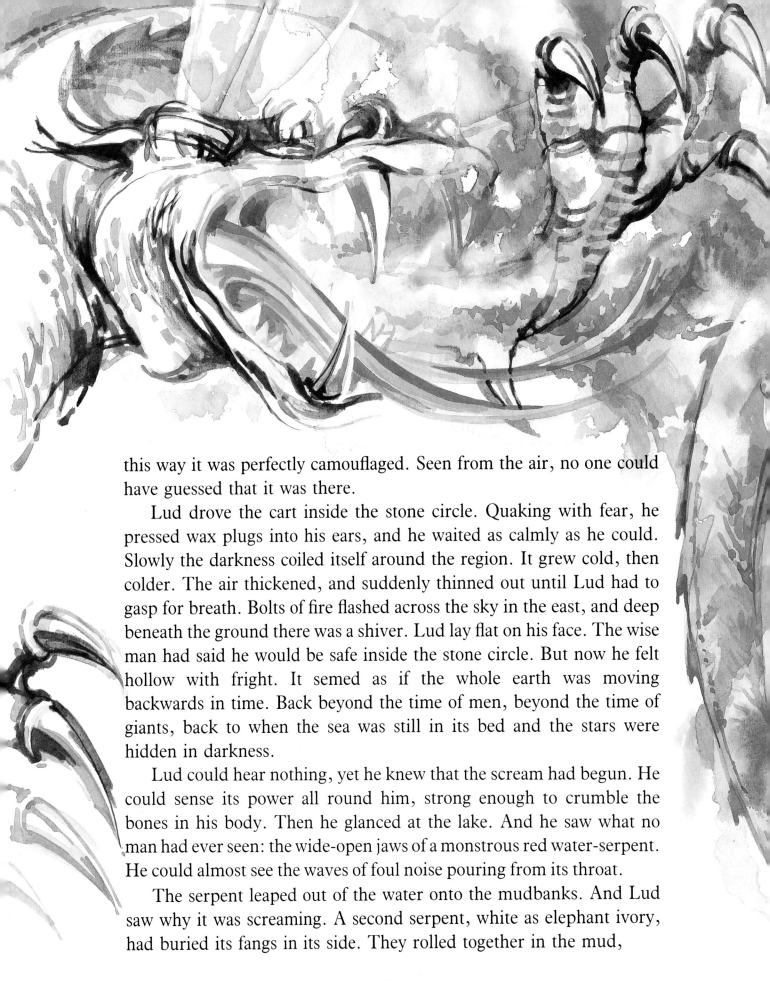

this way it was perfectly camouflaged. Seen from the air, no one could have guessed that it was there.

Lud drove the cart inside the stone circle. Quaking with fear, he pressed wax plugs into his ears, and he waited as calmly as he could. Slowly the darkness coiled itself around the region. It grew cold, then colder. The air thickened, and suddenly thinned out until Lud had to gasp for breath. Bolts of fire flashed across the sky in the east, and deep beneath the ground there was a shiver. Lud lay flat on his face. The wise man had said he would be safe inside the stone circle. But now he felt hollow with fright. It semed as if the whole earth was moving backwards in time. Back beyond the time of men, beyond the time of giants, back to when the sea was still in its bed and the stars were hidden in darkness.

Lud could hear nothing, yet he knew that the scream had begun. He could sense its power all round him, strong enough to crumble the bones in his body. Then he glanced at the lake. And he saw what no man had ever seen: the wide-open jaws of a monstrous red water-serpent. He could almost see the waves of foul noise pouring from its throat.

The serpent leaped out of the water onto the mudbanks. And Lud saw why it was screaming. A second serpent, white as elephant ivory, had buried its fangs in its side. They rolled together in the mud,

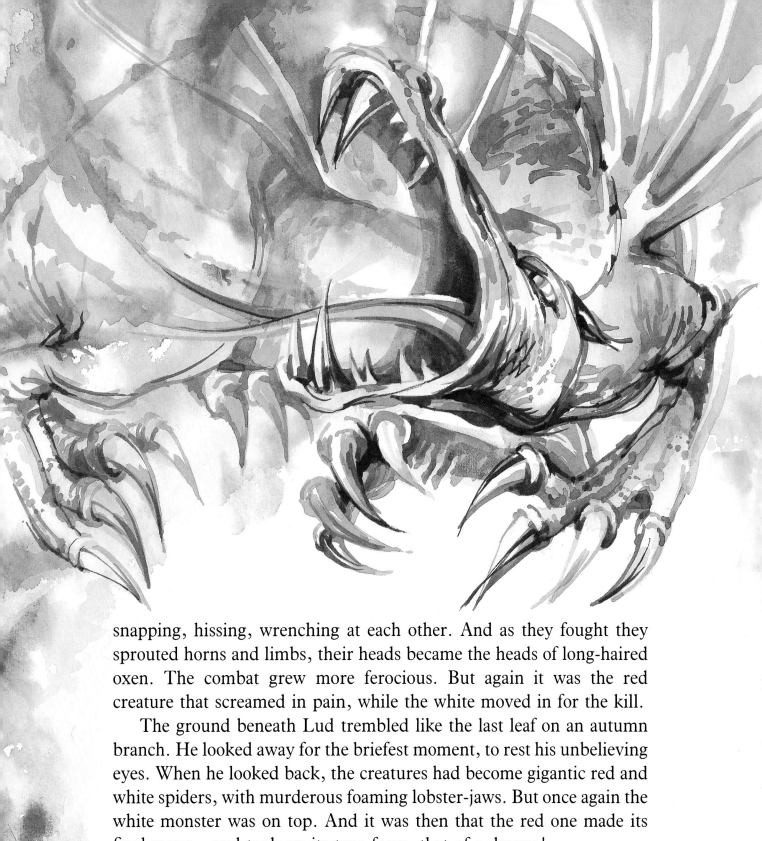

snapping, hissing, wrenching at each other. And as they fought they sprouted horns and limbs, their heads became the heads of long-haired oxen. The combat grew more ferocious. But again it was the red creature that screamed in pain, while the white moved in for the kill.

The ground beneath Lud trembled like the last leaf on an autumn branch. He looked away for the briefest moment, to rest his unbelieving eyes. When he looked back, the creatures had become gigantic red and white spiders, with murderous foaming lobster-jaws. But once again the white monster was on top. And it was then that the red one made its final move – and took on its true form, that of a dragon!

Up, up into the air it rose, in a flurry of blood and mud and pain. A winged red dragon, snorting shafts of yellow flame at its enemy, which turned at once into a white dragon, and renewed the conflict in the centre of the fiery sky.

But soon the light grew dimmer, the ground ceased to tremble so violently. Lud peered up and saw the dragons clinging together, all their energy spent. Soundlessly, gently, they were floating back down to the ground. As they came down, they changed at last into two little piglets, one red, one white, and both bleary-eyed with exhaustion. Down, down, down, gathering speed now, and before they could change back into water-serpents, they plummeted through the silken cloth and right into Lud's cauldron. Once inside, they guzzled up every last drop of the syrupy wine, and promptly fell into the soundest sleep. At that, Lud scrambled to his feet. He pulled the wax plugs from his ears and drove his cart to where the piglets slept. With nimble hands he fitted each one into a stone jar. When he had sealed the jars, he knotted the cloth around them, then loaded everything into the back of the cart.

Now he had to race against time. Before first light he had to drive high into Snowdonia, the snow-capped mountain range of north Wales. If he took any longer, the effects of the wine would wear off, and the creatures would awaken. But Lud drove his horses hard, and he arrived at the place called Dinas Emrys with time to spare.

It was a bleak, unwelcoming part of the Island, just under the shoulder of Mount Snowdon itself. And there, sparkling in the moonlight, was a small, four-sided lake. Without wasting a second, the young king gathered up all his strength and sent the two stone jars containing the piglets crashing through the surface. Immediately ripples began to fan out from the centre of the lake. Not ripples of water, but of earth, and small stones, and scrub and bristling grass. Before Lud had time to gasp, the lake had disappeared from sight! Now its waters lay deep inside Dinas Emrys, and so too did the fighting dragons. It would be a long, long time before the scream was heard again.

But Lud was still unable to rest. He headed back to London on a series of fresh horses. And all the while his thoughts were full of the food stealer – the third and last great plague of the Island.

Lud ordered a feast to be prepared for that evening. This surprised his cooks, for food was in dreadfully short supply, and now the young king wanted to use up every morsel in a needless feast. But Lud said to them, 'This food will not be for eating. I need it as bait.' The cooks did not understand what he meant, but they carried out all his instructions.

The feast was quite unlike any feast of old. The tables were piled high with mouthwatering food. But Lud forbade his people to lay a finger upon it. It was not a happy evening. On the table in front of Lud lay a sword, and beside him was the wise man's cauldron. This time it was filled with the coldest stream-water in the Island and a thick crust of ice had formed on its surface. Now and then Lud passed his hand across it, and watched his fingers turning purple. He knew what he had to do. And the knowledge was making his stomach churn.

Midnight came and went. Once again Lud placed wax plugs in his ears. He watched in wonder as, one by one, his people slumped forward in sleep. As the last man drooped, Lud's earplugs stopped having any effect. Then he heard the reason for the snoring all around him – the sweetest, most soothing, most delicate harp music ever played!

Lud strained his ears to hear more. But as he did so, drowsiness crept up on him. It tied weights to all his limbs and pressed hard against his eyelids. More than anything in the world he longed to sleep, to let this magnificent music flow through his dreams. But that was the last thing he could afford to do. He had to stay awake. He had to resist the magical music. This was the moment!

With a single sudden effort he twisted himself out of his seat. And he plunged headlong into the cauldron at his side! He smashed through the crust of ice on the water's surface, then disappeared from view.

But from whose view? The view of a clumsy, unshaven giant, who now shifted his huge frame into the feasting hall! He looked about him with his slow eyes. The sight of all the food made him begin to dribble uncontrollably. At once he crammed a whole cooked chicken into his mouth. And the wonderful music kept flowing from each and every pore of his skin. For this was the Mighty Man of Music. He lived in constant fear of being seen: as soon as any man set eyes on him, he would lose his musical power forever.

Then up crashed Lud through the icy water. He blinked twice, his drowsiness gone, every nerve in his young body tingling for action. He clapped eyes on the Mighty Man, and the music ceased with the suddenness of a thunderclap.

'Now I have you!' cried Lud, scrambling out of the cauldron and reaching for his sword. 'I'll make sure you do no more harm to this Island!'

The Mighty Man of Music dropped to one knee. Lud thought that he was offering himself for execution. It was all he deserved. But then the wretched giant spoke. 'King,' he mumbled through his beard, 'I beg you, show me mercy.'

'Mercy!' Lud laughed, coming closer. 'Why should I show you mercy? What mercy have you ever shown to the people of this Island?'

The Mighty Man looked tearfully into Lud's face. 'I am not to blame!' he moaned. The words were very unclear because he barely opened his mouth. 'My life has been one of unbroken misery. For hundreds of years I crouched alone in a corner of this Island, doing harm to no one. Then, from Anuvin, a curse was laid upon me. The curse of my musical power. I had to eat to feed the power. I needed the power to help me steal the food. Round and round and round it went! But now, thanks to you, my musical power is gone forever. I am harmless. Completely harmless.'

'So you are,' said Lud thoughtfully. 'And you look strong too. . .'

And so the young king decided to spare the Mighty Man of Music. He sent him to work at his court, and the giant proved to be a good and faithful follower.

That was how Lud saved the Island of the Mighty. But the men of those times knew so little. Lud had heard Bran the Blessed's voice – and he never realized it. He had seen the dragons fighting – and he never knew why. He had met the wise man of the woods – and he had not recognized him. The men of those times had so much still to learn. Soon they would understand the wise man's warning: 'If ever you see the third door, do not open it. . .' Soon, but not yet. Not yet. . . And in the meantime, the Island of the Mighty seemed safe.

The Dream of Maxen

THE Island of the Mighty had survived the three great plagues. But the realms of Anuvin would not leave this land in peace. Still they wanted vengeance for the destruction of Tara. So they turned to more cunning ways.

As soon as King Lud died, Anuvin laid its curse on his four sons. They became selfish raw men. They divided up the Island of the Mighty as if it were a hoard of treasure. Each son called himself a king. But they were nothing like the kings of older times. They cared little for their people. They were interested only in their own comfort and ease.

A cloud of unhappiness settled over the Island. The Islanders grew listless. The land itself seemed to sigh with sorrow. For when a land is ruled badly, something goes wrong with its heart and soul. In the depths of the White Mount, Bran's head understood what was happening. Bran knew that the sons of Lud had to be driven out. He knew that the Island had need of a true ruler, to bring new life, new hope. And he knew exactly who that ruler should be.

So what did Bran's head do?

It beamed a dream into the mind of a man. A fine man he was too.

46

He lived in the east, far from this Island's shores, in the world that we know. Maxen was his name, Maxen, the Prince of the Purple City . . .

Maxen was a wise and handsome Prince. He loved his people and his people loved him back. Although he was a fearless warrior, his greatest pleasure came from hunting. One day he gathered his noblemen together and took them on a stag hunt. From dawn to midday they chased their prey in a river valley close to the Purple City. Then the sun's great heat made Maxen drowsy. At once his noblemen made a canopy for him, by placing their shields on the shafts of their spears. Maxen made himself comfortable beneath it – and fell fast asleep.

When he awoke, he felt like a different man. For he had dreamed a most marvellous dream. A dream of a long, long journey. But all he could remember was the very last part of all. It went like this: He came before a girl. She was sitting in a narrow chair of gold. As soon as his eyes met hers, he had to look away. Her beauty was so dazzling, it shone brighter than the midday sun! She was dressed in flowing garments of white silk, fastened at the breast with gold clasps. Over these she wore a gold brocade surcoat, and a mantle fixed by a brooch of gold. But gorgeous as these clothes were, it was the girl's own beauty that made them sing. She rose from her chair to greet Maxen. He threw his arms around her. They both sat in the golden chair, which now,

surprisingly, seated the two of them with ease. Maxen's arms were around her neck. His cheek was next to hers . . .

But then, with the noise of the hunting dogs baying at their leashes, and the clashing of the shields overhead, and the stamping and whinnying of the horses, Prince Maxen had suddenly woken up!

In the week that followed, he lost all interest in the world around him. The life seemed to have been sucked out of him. Instead he was filled to overflowing with his love for the dream-girl. Not one bone-joint in his body, not even the middle of a single fingernail, was free from this love. And he could not bear to be parted from her.

When his noblemen went to drink wine and mead, he stayed behind. When they went to hear songs and be entertained, he stayed behind. He spoke to no one. All he wanted to do was sleep, so that his dreams might lead him back to the girl. And when he was awake he did nothing but brood.

'Where in the world can she be?' he asked himself over and over. 'Where do I begin to look for her?'

This was no way for a Prince of the Purple City to behave. His noblemen began to mutter among themselves. At the end of the week a chamberlain came boldly before him. 'Lord,' he said, 'Your men are speaking against you.'

'Is that so?' asked Maxen. 'And why should that be?'

'You neglect them, lord. You do not talk to them, or give them tasks to perform. They feel unwanted, useless, for they know that something is preying on your mind.'

'In that case,' sighed Maxen, 'I will tell you my problem. Last week, while I slept in the river valley, I had a dream. In this dream, I travelled far and came to a girl. If I do not see her again – in real life – I fear that I

might die. Yet I do not know where to find her, for all memory of my dream-journey has fallen from me.'

The chamberlain could hardly believe what he was hearing. But he said: 'Lord, ride out to the valley where you dreamed the dream that day. The rest of the dream might then come back to you. If it does, you will know which direction to travel in. And in the end you will either come to your girl – or find that she does not exist.'

'Oh,' said Maxen, smiling weakly, 'I am sure that she exists. I could never have simply *imagined* such beauty! But your plan is a good one.' So, for the first time in a week, Prince Maxen left his palace. His face had no colour, and he moved slowly under the burning sun. But at last he came to the river valley where he had slept. The chamberlain reined in his horse behind Maxen, pulling a face at the thirteen noblemen who had also come. All of them believed that this was nothing but a wild goose chase.

'This is the place,' Maxen declared. He gazed up the lush valley, his eyes became clouded, and suddenly a smile began to play around his lips. The details of his dream were crowding back into his mind! 'Yes!' he said in a soft, low voice, 'Yes! I began my journey by moving westwards, upstream.' Tears began to roll down his face. The chamberlain and the thirteen noblemen looked at one another, astonished.

'Yes, I remember, I remember!' Maxen continued. 'I kept travelling westwards until I came to the highest mountain in the world. I crossed the mountain, and found myself moving over the flattest, fairest lands. Strong rivers flowed down from the mountain to the distant sea. So I followed one of these rivers to its mouth.

'I came to a seaport city. Inside it there stood a huge castle, with many towers of different colours. Nearby a fleet of ships was at anchor. I noticed that one ship was far bigger and more splendid than the rest. As I drew closer, I saw a bridge of walrus ivory running between the quay and this ship. Before I knew what I was doing, I was crossing the bridge.

'The moment I stepped on board, a sail was hoisted, and I was wafted away over sea and ocean. It was a long voyage, and it ended at an island, the loveliest island I have ever seen. I set off to cross it, and eventually came to a harsh place of steep slopes and high crags. A river ran down from the mountains, and at its mouth was a mighty fortress, the most magnificent I ever saw. The gate to the fortress was open, so I went inside.

'The roof and doors of the hall seemed to be made of pure gold, its walls of glittering precious stones. There were golden couches and tables of silver. And there were people – the first people I had seen on my journey. On the couch facing me, two reddish-haired young men were playing chess. They were moving carved pieces of gold across a silver board. Their clothes were made of black brocaded silk. Around their heads they wore gold bands, studded with gems, and on their feet were gold-strapped shoes of the finest new leather.

'And that was not all. At the base of a hall pillar sat an old man. His chair was of ivory, decorated with two red-gold eagles. Everything about him was kinglike – his armlets, the rings on his fingers, the torque around his neck, the band around his head, all made of the purest gold. He was deep in concentration. In his hand was a rod of gold. And from it he was skilfully carving chess pieces.

'It was then that I looked beyond the youths, beyond the old man, and I saw my wondrous girl, seated in her narrow chair of gold . . .'

When Maxen had finished describing his dream-journey, he slumped forward in his saddle. He was exhausted by his yearning for this girl, and by the effort of all his remembering. 'That was more than just a dream,' he gasped. 'Truth lies wrapped in that dream. Truth! Now let us make the journey!'

'Lord,' said the chamberlain, 'You are not fit to make such a journey. Send these thirteen noblemen on from here. They will cover the ground as quickly as any men alive.'

Maxen longed to go himself. But he knew that his strength had left him. So he sent the noblemen in his place. 'Remember,' he called after them as they galloped away, 'tell her that she will be Princess of the Purple City. I will marry no one but her!' Then the chamberlain led him back to the palace, where he waited, and waited, and waited.

The thirteen noblemen rode on up the valley. The further they travelled, the less they joked about the Prince's madness. For here was the mountain as high as the heavens. Here the wide, flat plains, here the river. And there, at the river mouth, the seaport city with its multi-coloured castle. There was even a fleet waiting. Quiet with thought, they boarded the largest ship. And in that ship they sailed far across the Hazy Sea until they reached an island.

From the shore, the island looked lovely. But as the messengers rode across it, they saw that something was badly wrong. Both land and people seemed drained of all life. 'What do you call this country?' they asked one group of shuffling villagers.

'The Island of the Mighty,' they replied. 'But this is not the land it used to be.' The messengers shrugged at one another. They had never heard of such a place. They rode on, until they reached high Snowdonia in Wales. There they found the source of the river Seint. Eagerly they followed the river's course down to the coast. And finally, at Caernarfon, they came to the fortress of which Maxen had spoken.

The gate was open. The messengers dismounted and entered the

hall. Everything inside was just as Maxen had described it. They remembered how they had laughed behind their Prince's back, and they felt ashamed. The two young men did not look up from their game. The old man went on carving his chess pieces. But the thirteen messengers approached the girl in the golden chair and knelt at her feet. They kept their eyes on the ground. It was impossible for them to look hard at such blinding beauty. 'Greetings to you,' they cried together, 'Princess of the Purple City!'

At that the girl smiled. 'You seem to me to be noblemen,' she said, 'why then have you come here to make fun of me?'

'*Fun of you!* Lady, we do not understand!'

'You called me Princess of the Purple City – wherever that might be. What kind of joke is this?'

'Lady, no joke at all!' We speak as true as all the gold and silver in this hall. Our Prince, the great Maxen, saw you in his sleep. Since that day he has thought of nothing and no one but you.'

'In his *sleep*, you say?' asked the girl, smiling more broadly and raising her beautiful eyebrows.

'Lady yes! You must believe us. Come with us now, and be made Princess in the Purple City. Or, if you prefer, Prince Maxen will come here in person and take you for his wife. He swears that he will never marry another!'

The girl tossed her head in wonder. 'What a choice you lay before me!' she laughed. 'I do not doubt your honesty. But on the other hand, you say such extraordinary things!' She paused before going on. 'So you will have to accept this as my answer – if your Prince truly loves me, then let him come across the Hazy Sea and show himself to me. If he

52

pleases me, I will marry him. I can promise nothing more.'

The messengers rose to their feet, bowed, and made for the door. But as they were leaving the hall, one of them turned. 'Lady,' he said, 'can we at least tell our Prince your name?'

'You can,' the girl replied. 'My name is Helen. And this is my father, Cole. And these,' she pointed at the two young men, 'are my brothers Conan and Avon. Is there anything more you would like to know?'

'If I may be so bold, lady,' said the messenger, 'yes, there is. How is it that you live in such splendour here, even though your father does not rule this Island?'

The lovely Helen smiled, but it was a smile of sorrow. 'Your question is a good one. We are the last surviving members of this Island's greatest family. We are descended from Hu the Mighty, the first giant king. Bran the Blessed was our forefather too. But while Bran was away in Tara, his son Caradoc was imprisoned in this fortress. Ever since that time, no member of this family has ruled over the Island. But we have all been allowed to live in this marvellous place.'

'Do you never leave this fortress, then?' asked the messenger.

'The sons of Lud will not permit it. They have been warped by the curse of Anuvin. They let us do almost nothing now. My brothers here would make fine warriors, but they are forbidden to redden their swords with blood. My father would make the wisest of kings, but he is permitted only to carve chess pieces for some future game. I myself am old enough to marry, but no suitor has ever been allowed to enter this fortress. With every day that passes, the line of Hu and Bran moves closer to its end. And the Island of the Mighty is losing its will to live.'

53

The messengers kept these words in their heads as they sped back to the Purple City. They rode by day and by night, then went directly to Prince Maxen. Even before they spoke, Maxen rose from his seat. He could read the expressions on their faces. He knew that they had found his girl. But he listened closely to all that they told him. As they were talking, Maxen's old strength and spirit and wisdom returned to him. His noblemen and women saw this miraculous change in him, and it pleased them greatly.

He rewarded the messengers, then turned to the chamberlain, his eyes dark with determination. 'The waiting is over,' he said. 'It is time now to act.' And without further delay he summoned his army to him. That night, with the messengers serving as guides, Maxen and his troops set out for the Island of the Mighty.

The sons of Lud were not ready to face an invasion. They had no idea that Maxen's great army was on its way. When it arrived, they tried to raise armies of their own. But it was too late. The Islanders would not fight for such bad kings. They even welcomed the gentle invaders from the Purple City.

Maxen's troops fanned out quickly across the Island, taking control wherever they went. They drove the four sons of Lud into the sea, but they did no harm to any other Islander. No land had ever been conquered so smoothly, so bloodlessly. Maxen himself pressed on into Snowdonia. When he set eyes on the fortress at the mouth of the river Seint, he stopped and pointed. 'There,' he declared, 'is the place where I saw the lady I love best. It had to be more than a dream! I knew it!'

Minutes later he was entering the fortress and striding into the hall. He saw Conan and Avon at their game. He saw old Cole in the ivory chair, carving his chess pieces. And there, sitting before him in her narrow chair of gold, he saw Helen. Here, at last, was the girl who had shone in his dreams. 'Princess of the Purple City,' he said to her softly, 'I greet you.'

After a short time, Helen *did* agree to marry Maxen. How could she have done otherwise? But when Maxen asked her to choose her wedding gift, she replied at once, 'I should like something priceless,

something that you yourself have taken, but which does not really belong to you.'

'Name it,' said Maxen, puzzled.

'I should like the Island of the Mighty,' Helen told him. 'And I should like my father Cole to rule it in my name.'

Maxen agreed to this with delight. 'Now choose something for yourself,' he said.

'For myself, I should like three great strongholds to be built in places of my choosing. Then some good, straight highways should be laid down between them, so that men can travel easily through this Island. Now that my father is king, our land will recover its old strength. The strongholds and the highways will be signs of its new life.'

Maxen set his men to work. And when they had finished, Helen was more pleased with her present than with all the pretty trinkets in the world. And under the wise rule of old King Cole, the Island of the Mighty grew in health. The cloud of unhappiness melted into the skies. The fields clothed themselves in their old bright colours. It was a land to be proud of again, a land fit for heroes to live in.

Prince Maxen could not bring himself to depart. Instead, he set up his court at Caernarfon. And Helen had sacks of earth brought there from the Purple City, so that he could walk and sleep and sit upon his native soil! Many, many times the two of them spoke of Maxen's dream. They guessed that some magical power lay behind it. But they never found out that the power came from London, from deep inside the White Mount . . . Bran's head had foreseen everything. It had known that, in the end, the line of Hu the Mighty must rule the Island once again. And that was why it had sent the dream to Maxen.

For seven pleasant years the Island prospered. No one had ever known a time like it. A time of endless food and drink, of harmony between all men and women, of the most glorious birdsong from dawn to dusk.

'It must have been like this in the Golden Age,' Helen said to Maxen, 'when giants walked these hills and valleys.'

But then, at the start of the eighth year, a message arrived from across the Hazy Sea. It was delivered into the hands of Maxen, and it was as brief as a message could be. IF YOU COME, it said, AND IF EVER YOU COME TO THE PURPLE CITY. Maxen showed Helen the message. She looked at it in dismay. 'This is a challenge,' she said, 'a threat. And I fear that it comes from Anuvin. I beg you, lord, ignore it.'

But Maxen could not ignore it. 'My City is in danger,' he said, 'and besides, I have been in this Island too long.' Then he sent an even briefer message in reply to the threat: IF I GO TO THE PURPLE CITY, AND IF I GO. Within days he had gathered his warriors around him, and they made ready to go to the aid of their City.

Helen, troubled as she was, insisted on going with them. So too did her brothers, Conan and Avon. They led a large band of Island warriors, keen to win glory for themselves on the field of battle. It was almost as if they had all grown tired of the fruits of peace.

Old King Cole stood on the seashore. He watched the warriors streaming onto their waiting ships. He heard the excitement in their cries, he smelled their lust for blood – and tears began to roll down his face. 'I know that you will never return,' he said to Helen as he bade her farewell. 'When warbands leave this Island, they come to grief and perish. What has been will be again. But you will have sons. Send one of your line back here, to the Island of the Mighty. He will be a true king, a king of kings. Do you give me your word?'

Helen nodded her head firmly, then she kissed her father and tore herself away from him. With trumpets blaring, the fleet drew away from this Island's shores. Soon the ships were specks on the horizon. And then they were gone.

No one ever heard of that fleet again. Or the Purple City. Or Helen. Or Maxen. What happened to them? We cannot say; perhaps it is better not to dwell on their fate. But we do know what happened to the Island of the Mighty. We know that well enough. It is time, I fear, to tell the tale of Vortigern . . .

The Curse of the White Dragon

OLD King Cole reigned for many, many years. The Island of the Mighty almost ruled itself, as a kingdom will do when its king is true. Life became so sweet that time seemed to stand still. It was an age of laughter and forgetting, an age when black memories slipped from the mind – memories of wars and plagues, even memories of Anuvin. Eighty years passed, and they seemed like eighty minutes.

Then, quite suddenly, Cole's strength began to fail. As he lay on his deathbed, the Islanders muttered and furrowed their brows. 'Who will rule us now?' they asked. 'Who will wear the crown when Cole is dead?' The Princess Helen had promised to send one of her line back to the Island. But no one ever came. By day and by night the Islanders looked keenly out to sea, but no one ever came.

At last Cole summoned his noblemen to his bedside. Then he brought forth a set of golden chess pieces. 'Long ago I carved these from a rod of gold,' he said. 'One day a king will come, and he must have them. Until he comes, you must hold this Island. All of you. Not one of you must raise himself above the rest. And always remember this – if you should see the third door, do not open it. . .'

Those were his last words. But if he had known what was going to happen next, he would have died a second time. For one nobleman *did* raise himself above the rest. He stepped forward and he took the crown for himself. His name was Vortigern.

King Vortigern was not an evil man. But he lacked courage, he loved to be flattered and he was incurably curious. For several months, he moved his court restlessly from Caernarfon to Caerleon to Carmarthen. And all the while King Cole's last words were baffling him, *If you should see the third door, do not open it . . .*

He decided that he needed sea air, so he moved his court again. This time he went to Gwales, in Penfro. And there he found something unexpected – a splendid royal palace on the coast. Delighted by his find, he led his people inside. The hall was completely bare, but in three of the walls there were doors. Two were open. The third was closed.

At once Vortigern called for his Druids, his cleverest wizards. He pointed to the third door and said to them, 'Could this be the third door of which Cole spoke?'

'That is an exceptionally wise question,' replied the Druids, who took every chance to flatter the king. 'But in truth we do not know the answer. We feel, however, that it would be unwise to open it.'

'But why?' cried Vortigern. 'What can possibly lie beyond it? Only the sparkling sea, and then the land of Cornwall!'

'King, you know best,' the Druids said, their faces dark with alarm, 'But what can be gained from opening this door?'

Vortigern smiled back at them. 'What, though, can be lost?' he asked. His curiosity was simply too much for him. So
with that, he marched up to the third door
and pulled it back, laughing. . .

sunlight flooded the hall. It was just as Vortigern had said – there was nothing to see but the sea, and beyond that, Cornwall. The others gathered behind the king and looked out too.

And as they looked, a remarkable thing happened. All the years of forgetfulness came to a sudden end. Memory came crashing back to them like a single overwhelming wave. Every sorrow the Island had ever suffered, every tragic death, every cause for grief since the time of the giants – it all came surging back. And with such force! It was as if, in that one brief moment, they were experiencing every single one of the Island's past miseries – *on that very spot!*

They remembered Evil Efnisien. They remembered the gruesome war in Tara. They recalled the deaths of Branwen and Bran, the three great plagues of Lud's reign, the selfishness of Lud's four sons. And they knew once again of Anuvin and its curse. . .

'Close the door! Close the door!' wailed the courtiers, dropping to their knees in grief and fear. But there was no longer any door to close. There was no longer any hall. They were under the open sky, and the sea before them was teeming with countless ships, each one flying a White Dragon banner. The men on board were dressed in animal skins. Their faces, necks and arms were smeared with blood. And their shields were not raised upwards in peace.

Then one of their number leapt into the water. He was huge, the size of a giant, and he brandished a fearsome sword. Vortigern felt faint with terror, but he called out, 'Who are you? *What* are you?'

The giant roared back, 'You foolish, foolish man! I am Osla Big Knife. And these are my men – the People of the White Dragon. We come from the realms of Anuvin. Long have we waited for this moment! Long have we sought vengeance for the destruction of Tara. Now you have opened the third door and nothing can keep us out! The head of Bran the Blessed can help you no more. This Island will suffer as no land has ever suffered before!'

And already, inside the White Mount in London, Bran's eyes had closed. The ravens which had guarded his mighty head fluttered into the air and vanished. And in the fortress at Caernarfon, Cole's golden chess pieces vanished too. Bad, bad times were coming. Bad, bad times.

Then Osla Big Knife laid his great sword on the waters – and it grew

and grew until its point reached the land! Vortigern watched helplessly as the warriors of the White Dragon stepped onto the sword. But before they started their march towards him, he turned and fled, with his own people and the Druids close behind him.

'Run, run, foolish king!' Osla boomed after him. 'No matter where you run, we will find you. And when we do, you will die. Just as your Island will die!'

The Island of the Mighty had no defence against the People of the White Dragon. Time and again, they appeared in a sudden swirl of dust on the horizon. And after they had passed, there was nothing but smoke and waste and the howl of the wind through broken buildings. By night the skies glowed crimson, the land reeked of ashes and burning. By day the hills and valleys seemed to wail for mercy. Strong trees wilted, wild and dangerous magic flooded the air. The people knew that their land was crippled. But in their despair, all they could do was quarrel among themselves. And what of King Vortigern?

When he had fled far enough from Osla, he turned, breathless, to his Druids and said, 'Think hard. Where is the safest place in this Island to hide?'

'Wales,' they answered at once. 'We must make for the rugged mountains of the north, and build a fortress there.'

So they travelled high up into the crags and peaks of Snowdonia. On their way, they collected stonemasons, carpenters, blacksmiths – workmen they would need for building the fortress. 'The fortress shall be here,' Vortigern decided, pointing to the hilltop of Dinas Emrys, just under the shoulder of Mount Snowdon. It was a bleak place, smothered in brown scrub oak. The kind of place where men prefer not to be alone. 'I want my fortress walls standing by nightfall!'

When the men put down their tools that night, four great walls were standing tall, fifty feet above the ground. But Vortigern slept uneasily. He rose at dawn, and went to inspect the work by the light of a new day. And when he looked, he felt as if someone had scooped out his stomach. For not one stone, not one beam was there! Everything had gone – even the foundations. The fortress seemed to have melted into the hilltop!

Vortigern shook his Druids awake and demanded to know what had happened. 'King, we do not know,' they said. 'But build the walls again. Such a thing could never happen twice.'

By nightfall, the walls had one again been raised to fifty feet. By daybreak not a stone or beam was to be seen. 'Now what do we do?' Vortigern shouted at the Druids, 'Now what do we do?'

The Druids huddled together, and for several minutes they discussed the matter in low voices. At last they nodded gravely and said to Vortigern, 'A curse lies on this place. It can be lifted only by making a blood sacrifice.'

'I see,' said Vortigern. 'Then who is to be sacrificed?'

'A boy,' the Druids replied at once. 'A boy who was born of woman, yet never had an earthly father. You must find such a boy. We will sacrifice him here, then sprinkle his blood over the foundations to bind them together.'

So Vortigern sent out messengers to scour the Island for such a boy. It seemed to be a hopeless quest, in a land being ravaged by the People of the White Dragon. But as the messengers entered the town of Carmarthen, they saw two boys playing by the gate. Suddenly the boys began to quarrel, just as so many of the Island people did at that time.

'You cannot argue with me,' shouted the bigger boy at the smaller. '*I* have royal blood on both sides of my family. But *you* – no one knows where *you* came from. Your mother is a witch and you never even had a father!'

That was enough for the messengers. At once they sped the smaller boy back to the fortress site. They did not even stop to notice how strange the boy was, with his rotten teeth and his stink of hot iron. When they arrived on the hilltop, they handed him over to the Druids, who had set up a sacrificing stone. Vortigern stood some distance away, with his back to them all. He could not bear to watch the innocent child being killed.

Vortigern waited. And waited. And waited. But he heard no screaming. All he could hear was the wind, and now and then a thump against the ground. Cautiously he peered back over his shoulder through one half-closed eye. He cried out in disbelief at what he saw. For the boy was standing, unharmed, on the sacrificing stone, while below him there were the Druids – performing headstands, handstands and clumsy somersaults!

'What is the meaning of this?' Vortigern bellowed at them. And in reply, they did a most peculiar thing. They stopped their antics, stared at him, rubbed a finger across their lips and made a bubbling noise that went 'Blerwm, blerwm!'

Vortigern looked about him, bewildered. 'The boy has enchanted them,' whispered one of his people, 'He has used magic.' Vortigern took one furious step towards the boy. But when he noticed the boy's forehead, he stepped back again briskly. It was dome-shaped, far too big for the rest of his head. And it was *shining*! It was just as if the boy was beaming a radiant white light from inside his skull.

'What are you, boy?' Vortigern asked him hoarsely.

The boy did not look at him. His eyes misted over. His lips parted and an old man's voice thundered out from him:

'I have been many shapes
Before I came to be like this.
I have been a drop in the air.
I have been a shining star.
I have been a word in a book.
I have been a bridge for passing over
Three-score rivers.
I have journeyed as an eagle.
I have been a boat on the sea.
I have been with Bran and Lud.
I have been called Lassar,
Father of the fully-armed men.
I have been the wise man of the woods.
I have the knowledge of the stars,
Of stars before the earth was formed.
I have slept in a hundred islands;
I have dwelt in a hundred cities.
I have been teacher to all intelligences,
I am able to instruct the entire universe!
Idno and Heinin called me Merlin,
One day every king will call me Taliesin.'

Vortigern looked in confusion at his Druids. But they could only rub their lips again. 'Blerwm, blerwm! Blerwm, blerwm!' they said.

Vortigern raised his face to the sky. 'What am I to do?' he murmured in despair, 'What am I to do?'

'There is nothing you can do,' said the boy Merlin, looking straight at him now. 'You have done your damage by opening the third door. You can never be safe, you will always be sorry. Now tell your workmen to dig in the earth where your fortress has twice fallen. They will come to a buried lake.'

Vortigern gave the order to his workmen. They shovelled up a great mound of earth, and there, true enough, was the lake.

'Now tell them to drain the lake,' said Merlin. 'At the bottom they will find two stone jars, wrapped in a silken cloth.' The lake was drained. The bundle was found, and brought to Vortigern.

'Roll out the cloth,' Merlin told him, 'and empty the contents of the jars onto it!'

Everyone gathered round as the cloth was unfurled. Vortigern unsealed the jars, tipped them up, and out tumbled two sleeping dragons the size of piglets. One dragon was red, the other was white. At once they awoke, and began to fight. The white dragon seemed to be the stronger. It drove the red one to the edge of the cloth. But just when the red dragon seemed on the point of defeat, it rallied all its strength, turned, and drove the white dragon off the opposite edge. At that moment the cloth vanished. Then both dragons swelled up to their full awesome size. The white one rose high into the air, spitting fire. The red one rose in hot pursuit until they both disappeared from view.

For a short while, Vortigern and his followers remained silent. Then the king said humbly to Merlin, 'Explain to us, if you would, the meaning of what we have seen.'

Merlin stepped down from the stone. 'There will be a time of chaos,' he said. 'The People of the White Dragon will overrun this land from the eastern sea to the river Severn. The Island people will wish for death, but death will flee from them. It will be the darkest of dark ages.'

Then Merlin began to make his way down towards the sea.

Vortigern scrambled after him. 'Wait!' he cried out. 'In the fight that we saw, the red dragon triumphed in the end, and drove out the white one! Surely that means that the people will fight back and our Island will be spared?'

Merlin stopped and turned. Once more his eyes misted over, and the old man's voice thundered from him again:

'Alas for the old Red Dragon, for its end is drawing near.

Its mountains shall be levelled,

Its streams shall run with blood.

Twilight and the black of night shall move together.

But a new Island Dragon must come – a True King,

From far across the Hazy Sea,

Bearing four fragments of the Cauldron of Rebirth.

And the Cauldron must be remade,

Gently warmed by the breath of nine maidens.

It will not boil the food of a coward or a liar,

And it will be a sign of the land's rebirth.

The Island of the Mighty will be no more.

This land will be known by the name of Britain.

Then will the True King trample

The invaders' necks beneath his feet.

Then will the lands of the world

Be given to his power.

He will be loved and he will be feared.

His deeds will be as meat and drink to those

Who tell tales.

But first he must come

First he must come. . .'

Merlin turned and Vortigern reached out to him.

'When will he come?' he beseeched the boy, 'When will the time come for this Island to be saved?' But Merlin was no longer to be seen.

So the time of waiting began. The people of the Island went in fear and knew no peace. Always in retreat they were, like the snow that thaws upon the slopes of Mynydd Mawr. And slowly, with unbearable sadness, the Island itself stopped wanting to live.

What happened to Vortigern, the one who had opened the third door? Osla's troops ran him to ground near the river Wye. They tried to starve him out of his tower there. But soon they lost patience, and simply burnt both tower and king to black ash. It was then that Osla Big Knife issued his challenge:

'Men of this Island – if you call yourselves men – raise one more army to fight against me. I will be waiting, in the field by Baddon Hill. This will be the battle that matters. The battle to decide this Island's fate. Come – if you dare! I will be there!'

The challenge was heard all over the Island of the Mighty. Many Islanders were too idle to travel to Baddon. Others were too uncaring, or too frightened. This was truly a time of drooping men. But one good man set off at once, from that part of Wales known as Powys. His name was Rhonabwy. And a most astonishing journey lay ahead of him.

The Once and Future King

RHONABWY rode alone to the battlefield of Baddon. He rode through Powys but no man would go with him. He passed through broken villages and every door was closed. His heart ached when he saw the untended fields and the fear on the faces of children. Osla's army had turned the Island inside out. The smell of despair came strong on every wind.

Rhonabwy rode on. He left behind the lands that he knew, and a strange weariness came over him. He slumped forward in his saddle, but his horse took him on along paths that it chose for itself. When Rhonabwy raised his head at last, he recognized nothing in the landscape. He cursed himself, for he was hopelessly lost, and night was beginning to fall. Then, peering harder through the twilight, he noticed a single house on the road before him. It had a mean look, old and black, with a steep gable and smoke billowing through its roof. He went inside, and almost tripped on the bumpy, pitted floor. It was used by a herd of cattle. Rhonabwy tiptoed through mire to the far end of the hall. A tall, skinny hag sat there, tending a fire. Opposite her was a wooden platform, and on it lay a yellow ox-hide.

'I am travelling to Baddon and I have lost my way,' said Rhonabwy. 'I would be glad to spend this night here.'

The hag said nothing. But she threw a handful of chaff onto the fire, raising such a foul cloud of smoke that Rhonabwy had to withdraw into a back room. There he found a man, a saddlemaker. He was a sorry sight – bald, red and withered; and a strange smell came from him, a smell like hot iron.

'I am on my way to Baddon,' Rhonabwy began again, 'I would be glad to.'

'I care neither where you are going nor where you have been,' interrupted the man, then he spat. 'I know all that I need to know about you! Stay or go. Expect no hospitality from us.'

Rhonabwy thanked him. He was used to such bad manners. There was no courtesy or gentleness between strangers any more. It was all so different from the old times, so different. So many of the old ways were either dead or dying. Before Rhonabwy settled down to sleep, the hag brought him a wretched meal of barley bread, sour cheese and watery milk. Then, while a storm brewed outside, he stretched out on the dusty, flea-ridden scatters of straw. Rhonabwy tossed and turned, sick

to his bones. The Island seemed lost indeed, when hosts treated guests no better than stray dogs. 'When will the True King come?' Rhonabwy wondered aloud. 'When will he come and lift this land again?'

He could not sleep. So he rose from the floor, crossed to the platform, and laid himself down on the yellow ox-hide. It seemed to him that this might bring greater comfort. But as soon as his body touched the hide, he felt a shiver deep beneath the ground. He lay flat on his face, clutching at the hide. Then suddenly he was drifting, falling, spinning, tumbling and then – riding on horseback! Riding across a plain that he thought he knew! It was the plain of Argyngrog, and he was heading towards the River Severn!

Rhonabwy rode on. He hardly dared to think what had happened to him. Then he heard a horse and rider galloping behind him. He turned, and there in the distance he saw his pursuer – a fierce-faced man, brightly dressed in yellows and greens, with a gold-hilted sword at his side.

Rhonabwy took fright, and drove his own horse faster. But the fierce-faced man gave chase – and his was no ordinary stallion. For when it snorted out, Rhonabwy was blown far ahead. But when it breathed in, he was sucked right back to the level of its chest! Very soon Rhonabwy was overtaken. 'What is it that you want?' he cried, well used to being attacked for no reason on the roads. 'Show mercy!'

'Of course I'll show you mercy,' said the man, smiling now. 'What do you take me for? I merely wish to welcome you to Anuvin, and escort you to the camp up ahead at the ford.'

'Anuvin?' cried Rhonabwy. 'Anuvin! But this is the Island of the Mighty! I know this country. This is the wide plain of Argyngrog. Up ahead behind those trees is the River Severn!'

'You are right and you are wrong,' smiled the man. 'This *looks* like the Island of the Mighty. But in fact it is one of the realms of Anuvin. It lies beside and behind the world that you know. You entered this land when you touched the enchanted ox-hide. But do not be alarmed. No one here can harm you! Look hard at all you see, though. Things are not as they first might seem!'

'Then who are you, chieftain?' Rhonabwy asked the man.

'My name is Efnisien,' the man replied. 'Sometimes I am known as

the Churn of the Island, for crimes I committed in the far distant past. Now follow me, and I will lead you to the camp.'

'Are we to fight at Baddon then?' Rhonabwy asked, following.

'Wait and watch,' Efnisien called back over his shoulder. 'Wait and watch.'

They travelled on in silence until they came to the bottom of the great valley plain. Through the trees Rhonabwy could see a bright river. But he could also see a great deal more. For the road had begun to pass between the encampments of two great armies. Gorgeous tents and pavilions stood to both sides. The air was thick with the neighing of horses and the sharpening of swords. But the armies on either side of the road were very different from each other.

71

The warriors who thronged to the left were young and fresh, with scarcely a beard between them. The warriors to the right were fewer in number. Some of them were men wearing crowns, others were ancient Druids, others still were true giants, stained white with age. 'Look hard,' said Efnisien, pointing to the right, 'There is King Lud – a fine king in his day, though not a giant. And there is Cole – you must have heard tell of him? And there beyond them all is Hu Gadarn, Hu the Mighty – the first of the giant kings, the bringer of music and song. . .'

Rhonabwy marvelled at all he was seeing. But he did not understand what he saw. He followed Efnisien right up to a ford in the river, then drew alongside him. There in front of them was one of the strangest sights Rhonabwy had ever seen. On a flat islet in the water was the severed head of Bran the Blessed, its eyes closed tight, its mighty features sad with wisdom. And in the shadow of the head stood a man in splendid robes – a man who was every inch a king.

The King looked up, but his face was unclear in the shadow. 'Greetings,' said the King. 'You are welcome in this place.' Then he looked at Efnisien and asked, 'Where did you find this little man?'

'Yonder, my lord,' Efnisien replied, 'on the highway. He was on his way to Baddon.'

Then the King smiled a grim smile. 'My lord,' asked Efnisien, 'why do you smile?'

'Out of sadness,' the King replied, 'sadness that the Island of the Mighty should be guarded by such puny men, so different from the men of old. Go now, for the match is soon to begin.'

As they moved away, Efnisien said to Rhonabwy, 'Did you see the ring with the stone on the King's hand?'

'I did,' said Rhonabwy.

'Good, for that stone is magical. Now you will remember everything you see in this place. Had you not seen the stone, all memory would have fallen from you.'

At that moment a trumpet fanfare signalled that all should be silent.

'Let Cole's chess pieces be brought,' said the King, 'and let the match begin.' At that, a murmur of excitement passed around the watching warriors. A red-haired man brought the handsome gold pieces, and set them up on a board of silver.

'Bran,' said the King to the giant's head, 'Will you play chess with me?' And the eyes of Bran the Blessed opened.

'I will, lord,' came the answer in a rich voice, and at once the match began, with the red-haired man moving Bran's pieces to his instructions.

They were deep into their first game when a young page came running from a pavilion to the right of the road. He greeted Bran and said, 'The King's squires and servants are teasing your ravens and tormenting them. Do they do this with you permission? If not, then ask the King to forbid them.'

Bran did not take his eyes from the game. But he said to the King, 'Good sir, you have heard my page. Please call your men off my ravens.'

But the King only replied, 'Make your move', his eyes fixed on the board. And the page went back to his pavilion.

They finished their first game. Half-way through the second, another page came running to Bran. 'Lord,' he panted, 'the King's men are stabbing your ravens, killing some and wounding others. Surely this is against your will? Can you not ask the King to call them off?'

'Good sir,' Bran said gently to the King, 'if you please, call off your men.'

'Your move,' said the King. And the page went back to his pavilion.

At the start of the third game, a third page approached on horseback. His face was flushed with anger. 'Lord,' he cried to Bran, 'The noblest of the ravens have been killed. Those who still live lie wounded on the ground, barely able to flap their wings.'

'Good sir, I ask you again,' said Bran to the King. 'Call off your men.'

'Bran,' the King replied, 'it is your move.'

And with that, Bran turned his eyes on the page and he said, 'Go, raise the Banner of the Mighty where the fighting is fiercest. Then let the battle take its course!'

The page obeyed at once. As soon as they saw the ancient Banner, which had last been unfurled in Tara, the ravens soared into the air, seething with anger and a fearsome joy. The wind breathed new life into their wings. The course of the battle changed. Down swooped the ravens onto the heads of the King's men. They inflicted dreadful damage. Some carried off heads, some eyes, others ears and arms. The noise of their croaking and the men's screaming could be heard for miles around. At length, Rhonabwy saw one of the King's pages leave the fight, and gallop with his blood-stained spear towards the chess players.

'King!' cried the page, 'the ravens are killing your squires and servants.'

The King looked at Bran and said, 'Forbid your ravens to kill my men.'

'Good sir,' Bran replied levelly, 'it is your move now.' And the game went on.

Rhonabwy was fascinated, but perplexed, by all that he was seeing. He leaned closer to Efnisien, and whispered, 'What is happening here? What does all this mean?'

'The battle of Baddon is being woven here,' said Efnisien, deep in thought. 'All the battles of the future are being woven.' Then the skies were filled with the cries of men and the croaking of strong ravens. The ravens were dragging the men into the air, pulling them apart, and letting the pieces drop back to the ground.

A fifth page, heavily armed, came at last to the King and said, 'King, your servants and squires are dead. The sons of your noblemen will soon be wiped out. Without them, it will be hard to defend the Island of Britain when the time comes.'

'Bran,' said the King, 'call off your ravens.'

'Your move, good sir,' Bran replied, and they finished their game then began another. But the tumult from the battlefield became unbearable as the ravens ripped apart men and horses alike.

A sixth page approached and declared in fury, 'King, the ravens have killed all the men of your household, as well as the sons of your nobles. When the time comes to defend the Island of Britain, you will have nothing but your own might to call upon!'

On hearing those words, the King stood. With one hand he swept up the gold chess pieces, and squeezed them till they were nothing but a mound of dust on the silver board. 'Bran,' he said calmly, 'now will you call off your ravens?'

Bran's face smiled a knowing smile. 'Lord, I will, for I have seen enough. You are indeed the True King, come from across the Hazy Sea. I greet you as one of my line. The Island of the Mighty is no more. But through you the land will be reborn. And that land will be known as Britain!'

Then Bran ordered the Banner to be lowered, and it was brought and laid before the King. 'The Banner is yours,' said Bran. 'Now the giants and the good men of the past must depart. We will return to the Summer Country. As for my ravens, you may do with them as you please. I wish you well. I wish your Island well.' Then Bran the Blessed's eyes closed, never to open again.

Immediately a handsome young warrior stepped down onto the

islet. He knelt before the King, and presented him with the sword of victory. It was a magnificent weapon. The design on the hilt was of two gold serpents. And when the sword was unsheathed, two flames of fire seemed to shoot from the serpents' mouths. The brightness made Rhonabwy shield his eyes with his hand.

When he looked again, two more warriors were bringing to the King a great golden throne. It was large enough to seat three armed men in comfort. A final warrior produced a ribbed brocade mantle. He spread the mantle on the ground, so that all could see the embroidered red-gold apple at each corner. Then all three of them set the throne upon it, and the King took his rightful place upon the throne.

Rhonabwy turned to Efnisien. 'Tell me now,' he pleaded, 'what has been happening here?'

'What you have seen,' replied Efnisien with a smile, 'was the testing of the True King. The battle you witnessed was a rare one indeed. The two encampments were those of the forces of the Island's past, and the forces of the Island's future. Here in Anuvin, the web of the future is woven. All the King's battles remain to be fought. But because of what happened here today, all his battles will be victories. The Banner of the Mighty has now passed to him. He will not fail. He cannot fail.'

'But what will happen now?' asked Rhonabwy. 'Will the King ride to the Battle of Baddon?'

'He will, but not yet. First he must lead his men across the sea to Tara, to the fortress built for Bran. And he will bring back the four fragments of the Cauldron of Rebirth.'

'I wish to sail with the True King!' cried Rhonabwy. But Efnisien was guiding him back to the side of the road, where a yellow ox-hide lay stretched on the grass.

'You have seen what you have seen,' Efnisien told him, 'and that must be enough. You are a good man. You belong in the new Island of Britain, not in the realms of Anuvin. But at least you now know that the True King will come, that the Island will rise again.'

With that, he indicated that Rhonabwy should step onto the ox-hide. Rhonabwy took one last look around him, then took the step. . . Once again Rhonabwy was lying on the ox-hide. And he thought he had dreamed a marvellous dream. But when he opened his eyes, he saw that he was not in the filthy home of the saddlemaker. Instead he was inside a spotless but bare hall, the hall of a royal palace. There were doors in three of the walls, and all of them were open. Through the third doorway Rhonabwy could see the sea. And leaning against the opened door, laughing, was the saddle-maker.

'I dreamed,' Rhonabwy said to him, rubbing his eyes. 'Oh, how I dreamed!' But the saddlemaker continued to laugh, said nothing, then

beckoned Rhonabwy to the doorway. Rhonabwy rose to his feet and stepped forward.

There, through the open doorway, he saw a mighty ship of blue-green glass, thronged with eager warriors. It was heading out to sea, and above its ranks of men there fluttered the Banner of the Mighty!

'What you saw was no dream,' said the saddlemaker, behind Rhonabwy.

Rhonabwy turned, but the saddlemaker was gone. The hall was gone. 'Where are you?' Rhonabwy cried at the open sky. 'Who are you?'

And a voice came back from the thinness of the air:

'I am Lassar, I am the wise man of the woods.

Idno and Heinin called me Merlin.

One day every king will call me Taliesin.

This has been the Island of the Mighty.

Already it is being reborn,

And it shall be known as Britain.

This land will rise again,

Armed, but not in the old way.

The True King will come.

Watch and wait for he will come.

And his name will be Arthur!'

'Arthur!' whispered Rhonabwy, turning to see the ship of glass, no more than a speck on the horizon now. 'King Arthur! Come to us soon. come back to us soon.'

Oxford University Press, Walton Street, Oxford OX2 6DP

Oxford New York Toronto
Delhi Bombay Calcutta Madras Karachi
Kuala Lumpur Singapore Hong Kong Tokyo
Nairobi Dar es Salaam Cape Town
Melbourne Aukland

and associated companies in
Beirut Berlin Ibadan Nicosia

Oxford is a trade mark of Oxford University Press
Illustrations © Anthea Toorchen 1987

Text © Haydn Middleton 1987

setting, arrangement and editorial matter

© Oxford University Press

First published 1987

British Library Cataloguing in Publication Data
Middleton, Haydn
Island of the mighty.—(Oxford myths and legends)
1. Mythology,—Juvenile literature
I. Title   II. Toorchen, Anthea
293'.13   PZ8

ISBN 0-19-274133-0

Phototypeset by Tradespools Limited, Frome, Somerset.

Printed in Hong Kong